A SHORT
HISTORY OF
EVOLUTION

A SHORT
HISTORY OF
EVOLUTION

Carl S. Coon

HUMANIST PRESS

WASHINGTON, DC

Copyright © 2014 by Carl S. Coon

HUMANIST PRESS

Humanist Press
1777 T Street NW
Washington, DC, 20009
(202) 238-9088

www.humanistpress.com

Printed book ISBN: 9780931779558
Ebook ISBN: 9780931779541

Editors: Luis Granados, Matthew Bulger
Cover design: Lisa Zangerl

TABLE OF CONTENTS

INTRODUCTION

A few years ago I was at a meeting with Salman Rushdie and someone asked him if he wrote an outline before he started to write a novel. He looked surprised and said, "Why no, I never know how a book is going to end when I start it."

I was intrigued by this, having been brought up in the "outline first" school. I had written two books previously, and to the best of my recollection I had started each of them with a pretty clear idea of where I was going to end up.

I wanted this narrative to be different. I started by ignoring endings and searching only for the proper beginning. It was like composing music in the form of theme and variations. You start with a concept and if it's a good one things fall into place and you end up with all the variety you want built on an underlying unity, the theme that you started with. Beethoven's many exercises in the theme and variations pattern sound like they follow that organizing principle. And they make great music.

So does life itself. The trick, of course, is finding the right starting theme. I suppose that for much of my life I have been looking for the right organizing principle to serve as the platform for my evolving ideas about the big issues of what we are and how we got here. So, I suspect, have most people who bother to think about these questions.

It was when I reflected on entropy and saw evolution as the opposite that I realized I might be on to something. Here was a theme that began as an organizing principle and ended up fitting the facts I already had in mind. Could it provide the starting point for a science-based narrative that explains life on our planet, from its origin to our present human condition?

Why don't we already have such a narrative, simple enough to become conventional wisdom, like Genesis and other old origin myths, yet based on what we know rather than our fantasies?

Part of the problem used to be that until fairly recently there were too many gaps in what we knew, too many missing links in our attempts to construct a logical chain of evolutionary events. How to explain the origin of life, or the apparently divine spark that invested our own species with such a singular intelligence? As long as we had to fantasize when filling in these and other gaps, our argument was vulnerable to attack by creationists and others convinced that science would never find all the answers, and at the end of the day we would always find mysteries explainable only by divine intervention.

A more current problem is that we now have almost too much information. After all, the world's population has more than doubled during my lifetime, to over seven billion. That affects just about everything, including the number of bright scientists chipping away at the frontiers of what we know. They are not only much more numerous and much more specialized, but they also talk to each other, piggybacking on each other's discoveries like never before thanks to all the new ways we've devised to communicate. So we describe bits and pieces of the elephant and end up knowing an impressive amount about the beast, while still lacking a coherent picture of the whole.

What if we stand on the syntheses of others, try to put those parts together, and attempt to see the beast as a functioning whole? This will require us to simplify even further the simplified overviews of scores of talented writers who have already accomplished miracles of simplification in their own fields. In that second distillation, much will necessarily be left out, but if we see new patterns it will be worth it. It will be worth it even if we see patterns we already recognize, but in fresh perspectives.

Any sensible reader should stop at this point and ask, fine, but does the author know what he's talking about? Fair enough. Here are my qualifications, for what they are worth. I started life as the son of one of the last of the old-fashioned general anthropologists and grew up surrounded by talk about foreign cultures and human origins. I spent three and a half decades in the U.S. Foreign Service, back in the days when you could still get out and mingle with people and figure out for yourself what made them tick. After I retired, my wife and I traveled to many remote places we hadn't seen before, and I wrote a couple of books on cultural differences. Meanwhile I collected a whole bookcase of books dealing with the nature and origins of human behavior and related topics.

I'll list my main sources at the end of this study. Here, I'd like to single out a couple of authors whose works have been particularly helpful: Ian Morris, Peter Richerson, Christopher Stringer, Peter Turchin, and the two Wilsons, David Sloan and Edward O. Beyond books, Pete Richerson has given me invaluable direct advice and counsel.

At this point, it's time for the usual disclaimer. None of the sources I cite should be held responsible for anything I've said. I have played fast and loose with other people's ideas in my effort to construct a coherent overview of just about everything. I'm more interested in bringing together relatively unfamiliar principles and relationships, and seeing whether even less familiar patterns emerge, than in just adding another account of what has happened to an already overburdened bookshelf.

I'm happy to report that I've had quite a few of those "Aha!" moments of discovery while pulling my material together. I hope you, the reader, will too.

THEME
ENTROPY & EVOLUTION

Cosmologists tell us the universe began almost 14 billion years ago in a huge burst of energy, the Big Bang, and everything has been unfolding ever since following the general principle of entropy. After the Big Bang everything began to run down and energy to dissipate until finally, a few billion years from now, the whole grand show that started with a bang will end with a kind of whimper. It's not a feel-good kind of scenario, but it does give us a lot of time, and it lays the basis for some important principles.[1]

The idea of entropy has the kind of explanatory power we need to get back to first causes. It's one of the principles, like gravity, that is literally universal. You don't mess around with such concepts, but take them as given, solid anchors for whatever theoretical structures you want to erect over them.

But specifically, how do we define entropy? This poses a problem: all through this narrative I have had to consider how to define terms that annoy specialists when the public uses them in ways it understands, but are incomprehensible to the public when the specialists use them in ways they consider correct. Entropy is such a word. It has a specific use in thermodynamics and if you are a physicist that's fine, but I am not writing this just for physicists. For this study, entropy is diffusion of energy, a running down, a wearing out, and decay. It is what happens to a living thing after death.

Is there an opposing force that exists as the equal and opposite of entropy? If not, how do you explain life? I find it useful to use the word "evolution" in its broadest sense to provide a name for that opposing force. This implies that evolution isn't just something that emerged when life began; it was in-

---

1 The Columbia cosmologist, Brian Greene, has done a splendid job of explaining the Big Bang on the Ted TV series. Dark matter, string theory, and multiverses all enter into his explanation.

strumental in creating life in the first place. If there is life on other planets that life should be as subject to the basic laws of evolution as rocks on that planet are subject to the law of gravity.

If we can assume that evolution exists as a universal force that functions as the opposite of entropy, it follows that evolution creates complexity out of simplicity and concentrates energy.

These principles, complexity and power, appear in many guises in our narrative, but I hope to demonstrate that they provide an underlying unity for the whole evolutionary process on earth, just as Beethoven's initial theme unified the *Eroica* variations.

The test of any theory is how well it fits the known facts, and how well it predicts outcomes. One test of this theory will be whether the twin criteria of complexity out of simplicity and concentration of power fit what we know about both biological evolution and the quite different phenomena associated with cultural evolution. To the extent our theory works, we have a starting point for an analysis of evolution that finds comfortable places for both biological and cultural evolution. This should help clarify the longstanding controversies over gene-culture evolution, group selection theories, and sociobiology.

Evolution and Progress

Progress is a word that means many things to many people. For present purposes we define it as the direction followed by evolution. It relates to evolution the way decay and disintegration relate to entropy. It follows that progress happens when changes occur that move whatever is changing in the direction of concentration of power or greater complexity, or, usually, both.

On our planet, biological progress happens when entities reproduce in sufficient quantities to allow for a winnowing process, with selection favoring the ones that best fit the environment. Once an entity is created, there is no turning back, it will either prosper or die. If it survives and reproduces (or gets copied), it will become a platform for further evolutionary change, even if its introduction has thrown up new problems (which is likely to be the case).

Cooperation between entities is the engine of progress.[2] We sometimes call it symbiosis when it occurs naturally. Symbiosis can create a new kind of entity better equipped to cope with its environmental requirements. When that happens, the new entity will proliferate. It may even effect significant change in the environment itself.

Some changes can make the whole evolution game, as played in a particular environment, proceed on an altered basis. This transformation can be so gradual as to be imperceptible, or it can occur over a relatively short time span, like a step in a flight of stairs. If the changes follow the latter path, it can be a modest step, or something so radical and abrupt that it shakes up the whole environmental framework. When the change is sufficiently profound, we can construe it as a significant punctuation point in the whole history of evolution itself.[3]

This "short history" focuses on such punctuation points. It is not a narrative of what happened as much as an analysis of when and how and why the evolutionary process itself changed.

---

2 Nowak, Martin. *Supercooperators*. (Free Press, 2011.) See especially the last two pages of Chapter 13. Nowak uses a mathematical approach to study possibilities for symbiosis, or cooperation, between entities that can be as small as microbes or as big and complex as people. If his hope to reduce his equations to a simple formula bear fruit, we might someday have the equivalent of $E=mc^2$ for the basic evolutionary principle that is the opposite of entropy.

3 Coon, Carl. *One Planet, One People*. (Prometheus, 2004.) Pages 45-49, "Finding the Joints," elaborates on this thematic element.

FIRST VARIATION
THE ORIGIN OF LIFE

Before life on our planet began, parts of the environment favored the combination of molecules into increasingly large aggregations. Such aggregations might be helpful enough to the constituent parts so that they would stay together. This conjoined entity eventually might find other congenial entities that would fit, in the same sense, and produce a still larger entity. It was all a manifestation of the basic principles of evolution: complexity arising out of diversity, with the more complex entity possessing greater power than its constituent elements could muster if isolated.

Eventually, some of the more complex aggregations developed the capacity to reproduce themselves.[1] At first, it was a relatively simple matter of surviving and splitting. This produced very little variation, but there was still some. A few of these self-reproducing molecules happened to be better adapted than the competition to surviving, and their kind flourished. Meanwhile, molecular substructures evolved that added to the fitness and survivability of the structure to which they belonged. Eventually, single-celled organisms appeared on the scene and life as we know it took over.

Our telephoto lenses aren't strong enough for a close look at this period. As far as we can tell, there was no single path for the evolutionary process that produced life as we know it, no single point in either time or space when a critical juncture was reached. The gradient between the reactions we'd consider part of inorganic chemistry and those in what we now call organic chemistry is so flat that the argument as to where to draw the line appears more metaphysical than scientific.

We can infer that there were a lot of steps and maybe a game-changing breakthrough or two, not just one big leap from the inorganic to the organic. But that inference isn't proof, and many people still find it hard to believe that life could have evolved at all in the absence of some kind of intentional force, which some would call divine intervention.

I reject the whole idea of divine intervention on the grounds that it isn't needed. When you have a long enough time and are dealing with sufficiently huge numbers, that which has a small probability of happening at any given time and place can become very likely to happen someplace. When we are examining the period in which life began we are focusing on the molecular and cellular level and looking at eras that are billions of years in duration. There were many specific environments that could have existed during that unimaginably long period that could have presented opportunities for fe-

---

1 Humphries, Courtney. "Life's Beginnings." *Harvard*, Sept.-Oct. 2013.

licitous combinations. Once that odd combination occurred, if conditions favored its survival, its numbers might increase geometrically over generations, quickly changing something that started as a rarity into something quite commonplace.

Given the current pace of scientific discovery, we may well have better answers as to the origin of life in another generation or two. Meanwhile, we have enough evidence to give us confidence in our belief that life originated from natural rather than supernatural causes.

SECOND VARIATION
NATURAL SELECTION

1. Geospiza magnirostris.
3. Geospiza parvula.

2. Geospiza fortis.
4. Certhidea olivasea.

Darwin and the Definition of Life

Evolution as explained by Darwin is the modern equivalent of the Bible for those of us who wonder about how things got to be the way they are. His theory of natural selection explained so much about the life that abounds on our planet, and consequently proved useful in so many ways, that its essence soon became accepted as common sense in the more scientifically oriented elements of global society.

But even Darwin didn't explain everything, because he couldn't. He could not explain how characteristics were inherited because the world did not yet have genetics. Nor did he have cosmology, or population genetics, or other related sciences that have proliferated since his time. What he did give us was a path that led through the jungle of our ignorance to a place on the other side, as opposed to the paths others were proposing, paths that just left us milling around inside. What we need now is more paths like the one Darwin gave us that help us get closer to those mysteries on the other side.

Darwin identified natural selection as an evolutionary process that explains how living things came to be on our planet, without divine intervention, through descent with modifications. That is, living things reproduce themselves, producing copies that are not exactly identical to themselves. The copies that are best adapted to the environment are the ones most likely to survive and reproduce. Over time, this can produce significant changes in the form and the behavior of the life form. The process looks purposeful, but in reality it is no more guided by some overarching intelligence than the law of gravity.

Natural selection is a complicated theory, especially when we examine the copying mechanisms and the ways the environment and the life form can change each other. But its essence is simple enough so that the concept of life itself can usefully be defined in its terms. That which is subject to natural selection we can consider alive, while everything else is not.[1]

Natural selection is an evolutionary, anti-entropic process in that it concentrates energy and produces complexity out of diversity, but it is not the only such process. Human intelligence has produced a new environment within which a different anti-entropic process exists, which is known as cultural evolution. I'll get to that later.

---

1 *Cf.* Smith, John Maynard and Szathmary, Eors. *The Origins of Life.* (Oxford University Press, 1999.) My definition of life as that which is produced by natural selection is a simplified version of theirs.

Natural Selection at the Cellular Level

Single-celled organisms generally reproduce when individual cells reach a certain size and split, each half replicating the parent.[2] If the environment is favorable and each half has good prospects for growing up and splitting, then pretty soon the environment is full of them. Once in a while, minor changes occur within individual units during the splitting process, which can be called copying errors. Other changes, even less frequent, may be caused by external forces like radiation, which can be called mutations. Most of these changes have an adverse effect on that cell's odds of surviving. Those changes may survive for a few generations if the environment is favorable enough to tolerate their departure from the norm, but the organisms carrying them (or rather, their descendants) eventually get weeded out of the population to the extent they are competing for resources with other, better adapted cells.

Rarely, some changes will occur that favor the organism's survival. In simpler times, survival depended on the principle of "eat or be eaten," so any modification of the organism's structure that enhanced its capacity to sniff out the neighborhood and distinguish between food and foe would likely survive over many generations, and might even spread through the entire population. Such adaptively advantageous changes are the catalysts that make evolution through natural selection possible. They may be so rare that it is hard to conceive of their happening at all. But even if a change only happens to one individual in a million in its particular neighborhood, consider the possibilities if there are a million neighborhoods in the same pool, continuing on for a million generations. The odds are good that eventually enough of the lucky mutants will survive to allow natural selection to cut in.

As a general rule, evolution through selection speeds up

---

2 Dawkins, Richard. *The Selfish Gene.* (Oxford University Press, 2006). This book opens with an admirably detailed but lucid explanation of how life evolved during this phase.

when the environment changes in ways that put unfamiliar stresses on the population. The new environment may favor copying errors or mutations that are just emerging, or it may make lucky winners out of a minority that already had changes that either had been mildly advantageous or that were latent.

The causes of change in the environment can come on slowly, like an ice age, or abruptly, like a volcanic eruption. The change itself can be generated by life within the ecosystem, if for example a new predator or a new food source appears on the scene. Or the change can be self-generated, if for example a species is so successful it outgrows its home turf and the pressure of its own population forces some individuals out into foreign turf. That can introduce new variables into the process if the new environment differs from the old one, or if the emigrants are not entirely representative of the home population.

If there is no outlet for surplus population, and the environment remains stable, an optimum population will eventually be reached, establishing a durable equilibrium.

Sexual Reproduction[3]

Very rarely, something happens that changes the basic parameters of the evolutionary process. Sexual reproduction was such a change. It altered the basis for the copying function that enabled the basic cycle of birth and death punctuated by reproduction that is the essence of life as we know it. It did this by transforming reproduction from a relatively straightforward matter of copying to something more like a crap shoot. In the latter, you start with two identical objects, but when you roll them you don't know whether you'll end up

---

3 For simplicity's sake I have omitted several major breakthroughs involved in the evolution of sexual reproduction. These included "...the development of chromosomes from independent replicators, the emergence of eukaryotic cells from prokaryotic cells, the evolution of multicellular organisms from unicellular organisms, and the development of eusocial colonies." (Turchin, Peter *et al.* "A Historical Database of Sociocultural Evolution." *Cliodynamics*, Vol.3, Issue 2 (2012)).

with snake eyes or box cars or some number in between. In sexual reproduction each parent's characteristics get mixed together, and one never knows in advance how the offspring will turn out, except that whatever can be identified as a genetically inherited feature will be attributable to one or the other.

When you roll a pair of dice the number of possible combinations is limited. When living organisms mate the range of possible outcomes is several orders of magnitude larger, because the complexities of the DNA code in each individual vastly exceed the simple six possibilities engraved on a single die.   Sexual reproduction was a vast improvement over reproduction by fission from an evolutionary point of view because it greatly increased the possibilities for useful variations to emerge through natural selection. It provided a quantum jump in the odds that any given life form would respond adaptively to changes in the environment. It is certainly one of the most important innovations in the long history of the evolution of life on earth.

Once sexual reproduction had taken hold in our planet, natural selection came into its own as the primary engine of evolution. It ran amok, producing the huge archipelago of life forms that ever existed, including those that exist to this day.

Our narrative will skip rapidly through this period, because until we entered the scene, natural selection was really the only game in town. We'll note a few of the more important evolutionary developments and let it go at that, before proceeding to the massive changes in the process itself that came with human culture.

Blips on the Screen

We know a lot about what exists today and we can see back, with some help from the earth sciences, quite a way into the ancestry of current flora and fauna. But there is an incredibly long period before that about which we know very little. Here's a quick reference to some of the scraps of knowledge that we do have.

There is geological evidence that the Earth had little or no oxygen for much of its existence. Oxygen levels then spiked about two billion years ago, and fluctuated after that. There are possible correlations here with the evolution of early life, but it's too early to say much more: it is a new field of inquiry, one that is just getting started.

Then there is the *Burgess Shale*, a rock formation in Canada that contains a bewildering variety of fossils created about 500 million years ago. Their sudden appearance in the geological record and their extraordinary variety supported a relatively new theory of "punctuated equilibrium,"[4] which held that evolution proceeded by sharply defined steps, rather than in a smoother, more linear gradient. Opponents of this theory continued to argue that more gradual change is the normal way life evolves.

If there is any merit in the present survey, it should help put that argument to rest. Evolutionary progress is always linear, but its tempo or rate of change can change, sometimes abruptly, for various reasons, notably when there are abrupt climatic shifts. Much of the argument can be reduced, therefore, to the semantic issue of when a step is not a step. We shall find out as we continue this narrative that as we approach what appears to be a sharply delineated boundary we see not a line but a zone, within which various intermediate processes are happening that make possible what we earlier saw as an abrupt change. It all depends on the time scale you are using.

Another major event was the *Permian holocaust*. About 65 million years ago, a giant meteor crashed into Yucatan and created something like that nuclear winter that doomsday types were predicting during the height of the Cold War.[5] The result: massive species extinctions, especially the dinosaurs. The whole top of the food chain was lopped off, and then

---

4 Gould, Stephen Jay. *Wonderful Life, the Burgess Shale and the Nature of History.* (W. W. Norton, 1989.)
5 University of Texas. "Experts Reaffirm Asteroid Impact Caused Mass Extinction." March 4, 2010.

some. Some of the smaller reptiles became birds, in an evolution that had already begun, while mammals took over on the ground. The mammals were little fellows, by comparison with their gigantic predecessors, but with a different design structure they were better equipped to survive in new and frequently harsher environments.

That brings us up to date, in a sense. I'll fast forward to the era of *Homo sapiens* once we have taken a quick look at Neanderthals and some other members of the *Homo* family whose evolutionary trajectories were still governed primarily by the rules of natural selection.

Ancestral Cousins

A great deal of work has been done in recent decades to establish human origins, but important issues remain unresolved, and the chart is still full of blanks remaining to be filled in. Here is a very brief (and necessarily controversial) summary of the famous "missing links" between our own species and its immediate predecessors.[6] We are still dealing with a process that is almost entirely biological, with natural selection dominating, and cultural selection still waiting for its curtain call.

Several million years ago a line ancestral to the modern chimpanzee split into three main components, chimp and bonobo on one side, and the genus *Homo* on the other. *Homo's* further evolution was linked to major changes in the climate and topography of eastern and southern Africa, when much of those regions changed from tropical forest to savannah. The chimps and bonobos stayed in the forests and evolved very slowly, since they were already well adapted to an environment that was also changing very slowly. But *Homo* was in an environment where, in order to survive, massive changes were

---

6 Stringer, Chris. *Lone Survivors.* (Times Books, 2012.) The serious literature devoted to exploring our ancestral roots as a species is extensive, and new studies appear almost weekly. I find *Lone Survivors* to be the best overall source in that it is comprehensive, authoritative, and current--at least for the moment. It covers both the experimental bipedal models and the transition from *Neanderthal* to *sapiens*.

required in both behavior and physical structure. The most obvious physical change, and the one most directly related to the changed environment, was to become primarily bipedal rather than arboreal. This transition was followed (or accompanied) by a cascade of other changes. Many of these changes involved bone structure, which paleo-anthropologists have studied in detail from skeletal evidence. Some other changes, like those involving use of stone implements, fire, and dietary habits, can also be inferred from the archeological evidence.

During this period a variety of experimental models emerged. They all walked upright, and experimented in somewhat different ways with necessary collateral changes, mainly in the head, pelvis, and feet. Some of these pilot projects were more robust than others. Bipedalism freed the hands and made it easier to manipulate objects. There was a trend toward bigger brains. We also know from the archeological evidence that at least some of the variants had fire and crude stone implements. Since they were all competing for much the same ecological niche, it became like a race, or lottery: one winner, many losers.

At least one of these variants evolved along lines that gave it an unusual capacity to adapt to new environments. About two million years ago, it spread out of Africa into the far reaches of the Eurasian continent. Over many generations its descendants evolved into Neanderthals (in Europe) and similar subspecies farther east (whom we refer to, collectively, as *Homo erectus*). These populations, sometimes called primitive humans, were organized in kin-based groups, and there is evidence that they were acquiring the rudiments of what we now know as culture. But they remained essentially conservative. Their armory of stone tools didn't change much over a very long period, whereas the sapient humans that succeeded them took less time to develop more sophisticated flint knives and a wide variety of other implements.

So what do we see when we survey what we know about the Neanderthals and their cousins? I suggest that we see a species that had evolved to the upper limits of its physical potential. They had big brains and were capable of living in cold as well as warm climates, but they were conservative, in that they adapted slowly, by comparison with the sapient humans that followed them. They were pressing on a ceiling, but had not yet figured out a way to break through it.

About 50,000 years ago a group of people that looked like us and had evolved over the previous hundred thousand years burst out of Africa and quickly overran a great deal of Eurasia, including Southeast Asia and Australia as well as Europe. The earlier populations, as far as we can tell, disappeared. *Homo sapiens* won the race.

THIRD VARIATION
DECONSTRUCTING THE
PROMETHEAN SPARK

What was it that made the difference, that so distinguished our sapient ancestors when they burst out of Africa that they spread all over the world? The Greeks attributed it to a gift from the gods, with Prometheus giving humankind fire. Can science give us a better answer?

Introducing Intentionality

Homo sapiens is the first (and so far the only) species that has exploited the quality of *intentionality*. What is intentionality? How should we define it?

Some scientists use the term "theory of the mind" for a similar concept. For present purposes let's agree to take both terms as including a capacity to think proactively, that is, to imagine things or happenings that do not presently exist, and make them happen. Likewise, intentionality implies a capacity to envision conditions or situations which would be harmful, even when they do not yet exist—and to take steps to avoid them, or at least moderate their impact when they do happen.

Why do we bother to distinguish between intentionality and intelligence? Why not consider our acquisition of intentionality as just another stage in the evolution of intelligence, rather than a game changer? Here's a common sense answer: many life forms exhibit some level of *intelligence*, but if any animal other than man had been able to acquire *intentionality* as we've just defined it, we'd know about it, assuming we were even around to care.

There's another answer that's more central to the present discussion. There may well be a broad analogy here with the distinction we drew between single-celled organisms that reproduce by splitting and their vastly more versatile successors that evolved when a few of them developed detailed copying via sexual reproduction. The successors succeeded because they could produce a much larger number of variations in each generation, which greatly increased the rate at which natural selection took place. A supercharged selective process took off, leading eventually to the diverse life forms that now dominate our planet.

Likewise, intentionality changes the way the whole system works. It breaks the iron straitjacket of natural selection that enables change, but only via selection from one generation to another. Creatures of the human mind can be conceived, made, revised over and over again, and replaced, all in the space of a single human span (although some ideas, like religion, may live much longer). They are not alive, at least by our definition of life, though they can pass through many of the same stages as living things. We need to keep this distinction in mind, when we think about how these creatures of the human mind change and evolve over time.

What is Intentionality?

The human brain consists of upwards of 100 billion neurons that interact with each other as they click on and off, forming a myriad of patterns that collectively constitute what we think of as thought. Think of the brain as a gas oven. When it sleeps only the pilot light stays on, but when it's activated many jets produce an array of flames. The flames themselves are insubstantial, as is our thought. It is the patterns that count.

Thinking of thinking in this way lets us conceptualize the human brain as something that evolved from a more primitive organ that had fewer neurons to work away at pattern creation. This tracks well with the archeological evidence that our ancestral cousins had smaller brain cases, and that there was a gradual increase in cranial capacity that broadly correlates with other archeological evidence for human evolution.

Thinking of thinking in this way also encourages us to examine the problems of the mind-brain relationship and human consciousness as observers rather than participants. We have grappled with these issues for centuries from inside the box, without making much progress. Maybe it will help to get outside the box, and look at the what and why of human consciousness from a purely evolutionary perspective.

Furthermore, if our subject is the evolution of our species as a whole, we can avoid the subjective tangles we get into when we try to compare that many-splendored thing, intelligence, both between individuals and between groups. This narrative, after all, is about evolution, not how smart we are. It is enough for our purposes to say that *as a species* we are a lot smarter, more adaptable, and more versatile than any other species on our planet.

It is a basic premise of this narrative that things don't just happen; they evolve out of other things. They evolve out of a winnowing process that is governed both by the nature of the environment and by variations in their own structure. We call this process natural selection when the subjects are living things, including ourselves.

As with the origin of life itself, we will probably never be able to point to a specific event and declare that that is when we first became human. Too many different things were happening in too many places and times. But we *can* define a broad period and a geographic region within which we passed what might be called our critical tipping point, and examine both the environmental circumstances and other factors for clues about what happened.

We are talking about a period covering roughly the first half of the last hundred thousand years. Neither the environmental circumstances nor the archeological evidence for that period is as well-known or understood compared to what we know about more recent periods. We are reasonably certain, however, that periodic climate changes were important in forcing the evolutionary pace. We'll cover this in Variation IV.

Intentionality didn't just evolve by itself. It was enabled, one might almost say begotten, by two other evolutionary tracks that operated symbiotically and synergistically to produce the kind of proactive intelligence we call intentionality. Those two factors are *language* and *altruism*.

Language

Natural selection can produce remarkable results. Bats can fly freely in pitch-dark caverns, "seeing" through their ears by echolocation. Every species that finds an ecological niche undergoes natural selection tailored toward a better fit with that niche, and often the result is some quality or bundle of qualities that is unique. *Homo sapiens* is no exception. Among our many innovations, language is the one that most clearly distinguishes us from all other life forms.

We can communicate with gestures and body language and simple sounds, and do so quite frequently, but so do other animals. There are limits, however, to the complexity of the messages that can be sent this way. When you can put these relatively simple sets of sounds into a framework that everyone around you takes for granted, the stage is set for more complicated messages. A communicable pattern forms that is intrinsic to the whole, rather than just being the sum of its parts. This can add an entire new dimension to the capacity to generate, transmit, and receive information.

Take the analogy of automobile license plates. Say each plate consists of just half a dozen numbers. If you were brought up to believe that each of the numbers conveyed a separate message and the order in which they were sequenced didn't matter, you'd see this mélange as six separate messages, each conveying a possibility of ten possible alternatives, for a total of sixty possible different outcomes. But when they are numbered sequentially and we perceive them as part of a system, the possibilities shoot up to a million. That's because of our assumption that there is a framework, which in this case happens to be the decimal system. That system isn't exactly hard-wired, but the infant *is* predisposed to learn systems that function that way early in life, and that predisposition appears to be part of its genetic inheritance.

Why Did Language Evolve?

Any adaptation survives because in the short term it helps the organism survive and multiply. This had to be the case with language as well; its revolutionary long-term implications were collateral benefits that unfolded only later.

Among the more proximate causes, some theories postulate that speech helped man the hunter and woman the gatherer garner food. Speech evolved when the fossil record shows increasing consumption of large mammals. That requires hunting in groups, the larger the better. Improved ways of communicating between individuals can be useful, both for planning hunts and for executing them. It can also help on the gathering side, in passing on information about where and when to find which edible vegetables, herbs, and so forth.

This doesn't mean that our language capability only started after the African breakout. It must have started earlier, but it fits the evidence to assume that the environment in Eurasian hunting grounds gave it a major stimulus. Flickers of intentionality in Africa during the preceding 50,000 years may well indicate the early emergence of the first efforts of the human brain to develop the extraordinary new capability of language.

People love to talk about themselves and gossip about others in their group. This human quality probably was present even in our first sapient ancestors and may well have been another powerful incentive to start talking.

If something seems to be worth doing, evolution will find ingenious ways to make it happen. One theory[1] holds that language co-evolved with the evolution of an innate human ability to use it. Linguistic structures governing vocal communication grew more complicated in ways that were user-friendly and adapted to the learning proclivities of children, even as succeeding generations of children grew more receptive to learning them.

Is There a Language Gene?

A half century ago, the linguist Noam Chomsky first identified what he considered to be an innate human capacity for grammar. He based this on the observable fact that very small children can pick up their first language quickly, despite its complexities, even though no other animal shows this capability, and even though the same child may show much less aptitude for rapid acquisition of other information. But where did this so-called "language gene" come from? Was it a singular biological event, a kind of mutation that came on very rapidly in terms of the generational tempo at which biological evolution proceeded? Or was it essentially another adaptation, more cultural than genetic?

The answer is probably somewhere in between. Language was a product of the natural world, an extraordinary adaptation that succeeded far beyond just helping its bearers survive and multiply in a changing environment. It was also an enabler, a midwife that made possible the birth of a whole new kind of evolutionary process. The history of evolution has many markers, and in a longer history than this there would be thousands, but by any standards this one was a winner.

---

1 Deacon, Terrence William. *The Symbolic Species*. (W. W. Norton, 1997), pp. 102-110.

There is plenty of evidence in the world around us that demonstrates the instinctive nature of a special language learning ability in children. The lamentable fact is that adults don't usually retain it, for it tapers off rapidly with the onset of adolescence. Some individuals retain traces of it into their first decade or two of adulthood.[2] A small minority seems specially gifted with a lifetime ability.

Our human capacity for speech may still be evolving. Within our own lifetimes, Apple introduced the user-friendly laptop, with revolutionary effects on the way the public at large came to handle large bundles of information. As anyone my age has learned, the kids took to the new techniques immediately and painlessly, while we oldsters had much more trouble adapting. There are too many times these days when I feel like an *erectus* whose fumblings are tolerated with barely disguised amusement by my more sapient descendants. A major portion of my confusion is linguistic.

The lucky few language savants in our midst may be harbingers of further genetic evolution in our ability to handle information. And then there are the few gifted people, often politicians, who have absolute recall of hundreds or even thousands of individuals, such that they not only remember their names after an absence of years, but also details about their family. Other rare individuals have absolute pitch in music, or can see four or five moves ahead in chess, or can see farther into mathematical mysteries than the rest of us. Are these singular developments flickering premonitions of what is to come?

If we look beyond our own species and see evolution in its broader context we can isolate the breakthrough on language as playing a role as central to the evolutionary history we are

---

2 I supervised the School of Language Studies at the State Department's Foreign Service Institute for several years in the late 1970's. I ran a survey at that time on how well various categories of officers succeeded in mastering hard languages. Fluency was achieved only by a minority and they were all among the younger cohort.

describing as was the much earlier newfound ability of multicellular organisms to reproduce sexually. Sex was a change that enabled life to branch out and proliferate and populate the planet with life as we know it. Language enabled our ancestors to break out of the pack, and change the world.

In both cases, the factors that allowed such a game-changing breakthrough can be described as something that allowed a quantum jump in the capacity for managing large quantities of information.

Was there some special evolutionary development that kick-started language on a new track? The answer is *altruism*, the quality of doing good for others even at some personal cost. Our remote ancestors learned to talk and learned to be altruistic toward other members of their group at about the same time, and it is doubtful that either could have succeeded without the other.

Does Altruism Contradict Natural Selection?

Living things want to stay alive, at least until they've had a chance to reproduce. This is an essential property of life. According to natural selection theory, when there is a resource that individuals share, each individual will try to get as much as it can. When there is nothing to inhibit such selfish behavior, the most competent individuals will do best, and succeeding generations will reflect that success. When the resource is limited, the winners take all and the rest are eliminated.

There are exceptional situations where altruism occurs in nature despite the appearance that it contradicts natural selection. It appears in social insects, including some ants and termites, where only one individual in the colony, the queen, can reproduce. This means that her numerous sterile progeny can best ensure the survival of the collectivity's genes if they support and defend the queen, even at sacrifice of their lives. The authors of *Star Trek* may have had this principle in mind when they created the Borg.

Another track has been altruism based on kinship. An animal can sacrifice its own interests if by so doing it enhances the prospects of another that shares much of the same genetic information. Mother bear may risk her life to save her cubs if it enhances the odds that her genes will survive, whether in her or in cubs that have half her genes. People today still feel compelled to act more altruistically toward close kin than toward others. We can ascribe this to "family values" or to nepotism or to something in between, but it is there, a fact arising out of our genetic inheritance.

The kind of altruism we need to explain here is the human ability to empathize and cooperate with humans other than close kin. Unlike the ants, and momma bear, this kind of altruism doesn't have any obvious way of reconciling itself with the natural selection principle. And yet, altruism that involves more than kinship has prevailed in groups of humans ever since recorded history began and earlier. It is the essential glue that enables us to cooperate with other humans. When and how did we acquire it?

We need to distinguish between altruistic practices that evolved naturally and those that evolved culturally. In our own times, the failure to make that distinction has resulted in a lot of needless argument.[3] If we get embroiled in that argument here, this "short" history will be a lot longer. It's especially important at this juncture, therefore, to define our terms and stick with those definitions.

Biological evolution is the framework within which life on our planet evolves.

Natural selection is the process that governs how living things evolve.

---

3 Segerstrale, Ulrica. *Defenders of the Truth.* (Oxford University Press, 2000). We refer here to the great debate over sociobiology that raged within the scientific community during the latter part of the twentieth century and still continues to misinform some scholars to this day. I have relegated the controversy, including this admirable reference, to a footnote because that's where it belongs.

Cultural evolution is an offshoot of biological evolution that has emerged with the human introduction of intentionality and language. It is primarily concerned with cultural artifacts. Like its parent, its general direction is toward concentration of power and creating complexity out of simplicity.

Group selection (including multilevel group selection or multilevel selection) is a subset of cultural evolution that refers specifically to evolutionary processes between groups.

The Tragedy of the Commons

When there is a resource that individuals share, but it is only renewable over time and gets consumed so fast it gets used up, the individuals all lose, and both winners and losers suffer. This is the so-called tragedy of the commons, where a resource (grazing land held in commons) became overgrazed because there was no way to inhibit individual shareholders from taking as much as they could get, even though restraint would benefit all. In other words, cooperation favored every individual, but absent some outside force to get them cooperating, it didn't happen.

This theme of the tragedy of the commons has been a core problem for our species ever since we became one, and its many variations still surface today. We constantly strive to devise restraints on individual selfishness that are strong enough to keep us enjoying the benefits of cooperation. Concepts of morality and justice evolved out of a sense that sometimes individuals had to go against their selfish interests if the group was to survive and prosper.

The Prisoner's Dilemma

This is a way to look at the tragedy of the commons that was cooked up at the RAND Corporation in 1950, and has been analyzed in great depth since then.[4] Part of its beauty lies in the fact that you can either play it with people or set it up as a

---

4 Nowak, *Ibid.*, pp. 5-6.

computer game that can play through any number of repetitions more or less instantly.

You start with a mixed population of people. You divide them into pairs such that no one is already acquainted with the person across from him. Everyone gets a card on which he or she votes either to cooperate or not. Voting is secret; the other person in the pair doesn't know you, hasn't talked to you, and has no prior knowledge of how you may be voting. The rules are that if both parties opt to cooperate, they each get a small reward. If one party votes to cooperate and the other one votes the other way, the cooperator loses his bet and the non-cooperator gets a reward greater than what he would get if they both cooperated. If they both opt against cooperation, nobody loses but nobody wins anything either.

If you have an arena filled with such pairs, or set up a computer program that games this situation, and repeat the contest over multiple rounds, you find that the cooperators lose out. The winning strategy for the individual is non-cooperation, even though everyone would end up better off if everyone cooperated. Altruism has failed and selfishness prevails, and it all works out about the way Richard Dawkins' selfish gene theory says it should. The tragedy of the commons revisited.

This is normally the way natural selection works. Progress occurs when selfish individuals combine for mutual advantage, not when individuals act against their own advantage. Momma bears and termite colonies are the exceptions, not the rule. But the whole picture changes when you introduce the human factors of intentionality and language.

If a player can identify the fellow he happens to be paired with and expects to have a further round with him at some point, he is more likely to vote for cooperation the first time around, in the hope that even if the partner doesn't cooperate the first time he may later. In later rounds, if he remembers how his adversary behaved before, he will likely defect when dealing with a defector while cooperating when the other fellow is a recognized cooperator too. A computer game that incorporates this factor of *reciprocity* will produce a variety of outcomes depending on how the odds are set.

If every player in the game knows how every other one voted in all the other person's encounters and not just with his, that reciprocity factor becomes strengthened to something like an absolute. Add this factor of *reputation* and the good guys will win every time.

Reciprocity and reputation thus form a solid basis for group cooperation in societies small enough so that everyone can at least recognize everyone else in the group. This is how societies were organized in the period when our sapient ancestors were hunting and gathering, and it carried on for a while after the introduction of agriculture. There is no mystery here. We all live with a limited number of people we know and can trust, including both kin and friends and acquaintances. Back when we all actually lived in villages, altruism prevailed effortlessly and naturally. It did not require any genetic modification—it was an effect intrinsic to our newly evolved mental capabilities. It co-evolved with language and culture.

The further evolution of altruism is an integral part of the complex of behavioral changes and technological innovations that our ancestors used to bootstrap themselves out of the village level into more complex societies.

FOURTH VARIATION
OUR EARLIEST HUMAN ANCESTORS

Now that we've had a theoretical discussion of the critically important transition from *Homo erectus* to *Homo sapiens*, let's get back to the narrative, and trace the early evolution of our own extraordinary species, *Homo sapiens*.

Out of Africa[1]

Homo sapiens' ancestors almost certainly lived in eastern and/or southern Africa, along with his bipedal cousins. There is some archeological evidence that humans with skulls shaped like ours go back over 100,000 years. Evidently the climate in southern and eastern Africa was relatively harsh during the first 50,000 years of that period. Our ancestral population remained small and isolated in pockets. There are suggestions of a rudimentary intentionality in a few ancient hearths, tools, and bones, plus rare traces of red ochre (possibly used

---

1 Stringer, *Ibid.* A good general reference for *sapiens* as well as *erectus* during this period.

symbolically for body decoration). The absence, so far, of a continuous chronology during this early period suggests that sapience may have flickered on briefly here and there and died out again, until finally, like a forest fire, it took over.

Around 50,000 years ago, climatic changes made the southern parts of Africa more hospitable. It seems reasonable that as the food supply improved and population increased, more people had the opportunity to leave their isolated pockets and mix with other groups, expanding cultural interchange. This may have produced surroundings enabling a few sparks to ignite a more widespread awakening.

If we can extrapolate back in time, we can infer that this awakening led to more food and, hence, to expanding population. This need not necessarily have taken a long time, as long as there was plenty of food.

Whatever the proximate causes, our sapient ancestors emigrated out of Africa in a couple of waves, or more, beginning about 50,000 years ago. The first, or one of the first, was a group that followed the coastline from Africa east all the way around to Australia. Remember that this period was during the last ice age, which tied up so much water in glaciers that sea levels were lower and coastlines were quite different. Other waves of these emigrants went east and northeast, and northwest to Europe. The Neanderthals disappeared, except for some intermixture with our ancestors (European DNA tests now show a small percentage contribution, less than five percent.)[2] The *erectus* cousins of the Neanderthals farther east suffered a similar fate, also leaving behind a small contribution to the new bloodlines there. Some of the easternmost ancestral humans crossed the land bridge to the Americas by about 12,000 BCE and spread from there to populate the New World.

---

2 Very recent research suggests that Neanderthal genetic material affecting hair and skin survives in modern humans in higher proportions than genes affecting other parts of the anatomy. Sankararaman, Sriram, et al. "The genomic landscape of Neanderthal ancestry in present-day humans." *Nature*, January 29, 2014.

How did these migrations differ from those of the more primitive peoples who preceded them? Intentionality based on language and altruism made all the difference. It enabled cooperation between larger social groups, with all the accompanying benefits size confers, in terms of division of labor, ability to survive in bad times, and the ability to prevail in competition with other groups.

Our ancestors developed new and more efficient kits of stone tools during this period, as shown by the archeological record. DNA evidence (of body lice, of all things!) shows that they learned early on how to fabricate warm clothing, permitting bodies that evolved in warm regions to capitalize on abundant food supplies in the colder northern climes. And we also have signs of early culture, including ornaments, flutes, and of course the famous cave paintings. All this newly minted adaptability allowed the newly minted people to eat better and to move much more rapidly, both across space and through time, than their *erectus* predecessors

Natural selection, as we've already established, involves changes punctuated by generations. The length of a generation didn't change much when people became sapient, though much else did. When *erectus* burst out of Africa about 200,000 years ago, they settled in places where the weather and land forms were very different from the home turf where they had evolved. However, the tempo at which they spread out was slow, spaced over many generations, so adaptive pressures on each generation were not as strong. When *Homo sapiens* emigrated much later, the tempo was faster. The new intentionality factor started a positive feedback loop; increased adaptability enabled more rapid dispersion into new environments, and the challenges of adapting to the new environments tilted the selection process toward favoring the most adaptable individuals.

This feedback loop applies in principle to many other behavioral and physical features that distinguish *Homo sapiens* from Neanderthal and other cousins, so it is reasonable to regard this period, from 50,000 till 10,000 BCE, as a rich period for rapid change.

It is illustrative to turn this tempo change around and apply it to the environment instead of the invaders. Various species of large herbivores that *Homo* relied on for food during both invasions survived the first invasion, but during the second one, in an eerie preview of coming attractions, they went extinct (the wooly mammoth, for example). Our sapient ancestors had become too efficient as hunters and too disruptive of the natural balance for the existing equilibrium to survive its introduction. Great herds of herbivores survive to this day in the African heartland because they co-evolved with these strange new bipedal carnivores; there were enough generations while *Homo* was growing up to permit them to grow up too, through natural selection.

Social Organization

Because the climate was harsh and fluctuated rapidly, humans were scattered thinly across Eurasia. The food supply for our hunter-gatherer ancestors wasn't predictable enough to support denser populations. People probably arranged themselves through space in ways somewhat analogous to the ways they do now in remote mountainous areas, where inhabited patches are separated by mountainous terrain. If that analogy has merit, then populations might be centered around nuclei of up to about 150 persons, which is the usual maximum for communities where everybody knows everybody else. Where food supplies were abundant and regular enough, such groups might coalesce into somewhat larger units.[3]

The more isolated these groups were, the sooner they might

---

3 There is a library of books dealing fictionally with that remote period. Jean Auel's *Clan of the Cave Bear* series, and William Golding's *The Inheritors* are classics in this genre. Auel is better on life with the Cro-Magnon, Golding is probably more accurate for the Neanderthal. Auel, Jean. *Clan of the Cave Bear*. (Random House, 2002.) Golding, William. *The Inheritors*. (Harcourt, Brace & World, 1962.)

develop dialects and other distinguishing features. Small neighboring groups might get together from time to time for the leaders to talk out matters of general concern while everybody else did business or celebrated. Most people would live and die within a few miles of where they were born.

When a group's total population would grow beyond the level the land could comfortably sustain it would split, with a part of it moving on to new turf. This raises the issue of genetic drift. In biology, generally, if an emigrant population finds itself in an environment that differs enough from that in the homeland it will evolve differently, and if it remains isolated from the parent group it may develop into a new species. Something along those lines happened to our own ancestors, accounting for present racial differences, but no speciation developed. Presumably the stresses of environmental change came too rapidly to be fully met by anatomical evolution, while our new sapient adaptability stepped in, as with warm clothing in northern climes, to help cope with the new stresses. Another factor was increased mobility, which apparently allowed enough gene flow across clinal borders to offset trends towards speciation.

Cultural evolution during this period opened new opportunities for a more complex division of labor. If the economic unit is no bigger than perhaps fifteen people, there's not much chance you can spare one of them to specialize exclusively in tool making, or basketry, or learning about the therapeutic qualities of naturally occurring herbs. With 150 in the unit, it's easier to spare someone who does nothing much but knap flint, since others who hunt and gather can feed him in exchange for his product. With intentionality this kind of specialization becomes not only possible but, eventually, necessary for survival.

With a minimum of coercion, this kind of group can maintain a kind of equilibrium which, at the time, was something new on the planet. Why? Not because it united many individuals into a cooperative whole—ants and termites can do that, and have done so for a very long time. The sociality of our sapient forebears was unique because its unity depended on the capacity of the individuals composing it for altruism based on intentionality and language. And unlike the bees and the ants, our new capacity for organizing contained the potential for further evolution.

A massive climate change came about 10,000 years ago and sparked the introduction of agriculture, which in turn kicked off further radical changes in the ways most people organized themselves in communities.

FIFTH VARIATION
THE NEOLITHIC ERA

Agriculture and Animal Husbandry

The Neolithic era began about 12,000 years ago, when our ancestors started to grow their food instead of hunting and gathering it. Cultural evolution began earlier, with the emergence of human intelligence (or intentionality), but was developing at a more stately tempo until agriculture and the domestication of animals came along and changed almost everything.

Where and When It Began

Settled communities based on agriculture and animal husbandry first appeared in hilly areas north of the Euphrates, now parts of northern Iraq and Syria currently occupied by Kurds. Archeological evidence of settled villages goes back in that area to about 9000 BCE. The combination of naturally occurring grains, and animals that were relatively easy to domesticate, was more favorable there than anywhere else in the inhabited world.

Farther east, other Neolithic settlements appeared a millennium or two later, starting in the Indus Basin and in central China. Each of these regions also harbored a combination of easily domesticated plants and animals that was almost as favorable as the first Garden of Eden in Iraq.[1]

The trigger that started it all was climate change. Around 10,000 BCE the weather changed dramatically for the better, so much so that we mark that period as the beginning of a new geological era, the Holocene. The planet warmed up, rainfall patterns settled down, and so did our ancestors. There's still some debate whether the new way of life evolved independently in each of the three areas or whether it spread from the first one. History and prehistory are studded with examples of cultural diffusion. However, the archeological evidence supports the idea that the three areas developed agriculture independently. For one thing, the suite of grains and animals that were first domesticated is rather different in each of the three areas.[2]

Once established, the new way of life spread rapidly. Some of the expansion involved absorption of resident hunter gatherer populations as the agriculturists took over their lands, but another common outcome was simple displacement, pushing the older populations back into less desirable turf. Our own takeover of North America from the existing Native American peoples was essentially a re-enactment of a drama that had been going on in many parts of the world for a dozen millennia. In virtually all cases, the odds were heavily tilted in favor of the agriculturalists, because there were generally a lot more of them when push came to shove.

---

1 Diamond, Jared. *Guns, Germs, and Steel.* (W. W. Norton, 1997). Excellent overall history of this era, for a general audience.
2 Richerson, Peter J., Boyd, Robert and Bettinger, Robert L. "Was Agriculture Impossible During the Pleistocene But Mandatory During the Holocene? A Climate Change Hypothesis." *American Antiquity* 66: 387-411 (2001).

Who Invented Agriculture?

Who first selected wild grains and domesticated them? We know that before the Neolithic era, people sometimes gathered the seeds of wild grains and used them for food. The problem with wild varieties is that their seeds ripen at different times so that they don't all get broadcast at once. So people (usually women) had to go through fields and shake the ripest pods, which was a time-consuming method. A clever person (again, likely a woman) started saving seeds from the relatively rare stalks that did all pop at once, and replanted them. It worked, and agriculture was born.

A Major Turning Point

If you think of evolution in terms of increasing complexity and the concentration of power, agriculture was a major turning point. In natural selection, the food supply normally operates as the most important single limit on population growth. Remove that limit and population can increase in a geometric progression. And it did with people when agriculture provided a reliable new source of food.

But that was only the start. A fertile region could support a much larger population by farming than by hunting and gathering, but in a few generations the population would multiply and something would have to give. When the Neolithic world was still young the most obvious answer when there wasn't enough land any more was to move on into fresh territory, and of course that happened and explains the rapid expansion of the Neolithic way of life throughout Eurasia and even into the New World. This is how biological evolution works. But in our case there was another dimension to the process, cultural evolution. The uniquely human capacity for intentional thinking enabled our forebears to adapt much more rapidly to new environments and new circumstances. And it enabled the people who stayed behind, when the young folk wandered off in search of new land, to grow and prosper as well.

Necessity is the Mother of Invention

Both the pioneers and the stay-at-homes faced many challenges as they went about colonizing the planet. Say the problem at hand was cycles of flood and drought that kept interrupting the food supply. One solution was granaries to store enough food to carry people through droughts. But any such adaptive changes would spawn new challenges requiring further adaptations. Humanity found itself on a kind of treadmill, with one thing leading to a couple of other things, an accelerating process seemingly without end.

If we want to know how we got on to that treadmill during the Neolithic era, we need to look beyond the immediate causes and into what happened to the basic nature of the evolutionary process itself. For starters, how does one characterize the "invention" of agriculture? It wasn't natural selection, it was intentional. It wasn't really cultural selection, nor was it a product of group selection. It seems to me that if there is a force or principle that operates to make an important evolutionary change happen, we have an example of it here. You don't need divine intervention, you need a convergence of factors that create a potential market for something that doesn't yet exist, at least in any visible form, and a convergence of existing entities that have the potential of getting together and meeting that emerging need. That convergence of factors, if strong enough, can vastly improve the prospects for a particular kind of variant in the system, making it not only viable but eminently desirable. And then that lucky variant may get discovered. It's as though the planet had a big computer, capable of finding a very small needle in a very big haystack, scanning for some as yet unknown thing.

Can that convergence of factors increase the odds that the lucky variant will be found? If so, can this kind of event help us understand how the human intentional mind got started? Could its principle also help explain the emergence of the first life out of our planet's primordial soup?

If agriculture did start independently in three different places, then it provides a kind of laboratory for testing theories of how major evolutionary breakthroughs emerge. We could look, for starters, for similarities and differences that might shed light on the relative importance of environmental factors. Is it too much to hope that such studies might prove useful for cancer research or conflict resolution, to take examples at opposite ends of our history of evolution?

SIXTH VARIATION
THE GODS OF WAR

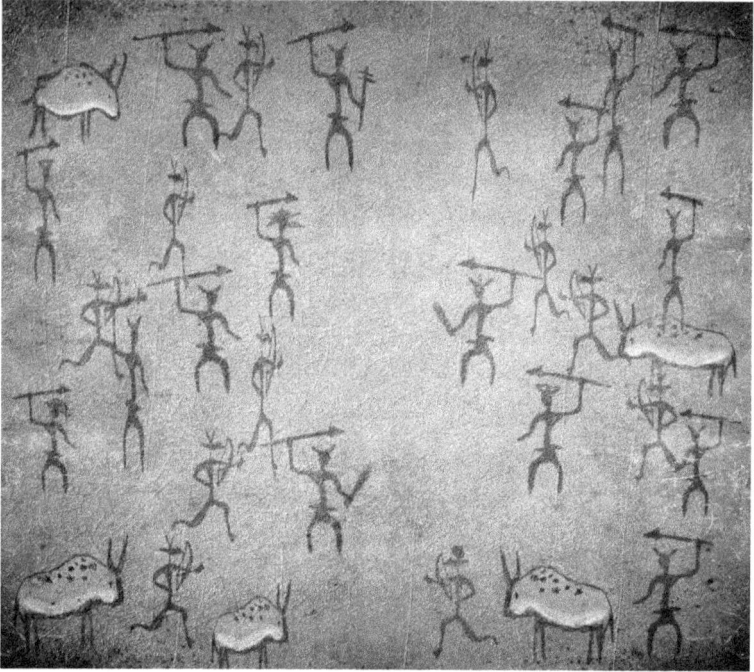

During the Neolithic era the rate of change, or progress as some people call it, sped up considerably. It was like a car shifting into a higher gear. But even so, village life was conservative by our standards. If it had taken our ancestors as long to get out of the Neolithic as it took their Upper Paleolithic ancestors to get into it, we might still be living in villages and using stone tools ourselves.

One thing led to another. When one problem was solved, several new ones would emerge in its wake. It was the same old principle, the simpler leading to the more complex, but like Beethoven's *Eroica* variations that same theme kept reappearing in very different forms.

It's not easy to distinguish between cause and effect here. It's like a stew on the fire that slowly begins simmering and then comes to a boil. Which ingredients had to come in first? Perhaps that is not as important as the fact that once the process started, they were all in the pot together.

The bottom line is that human intentionality combined with favorable environments to produce increasingly large groups of individuals who were willing to cooperate for the common good, even at some personal cost.

Let's take a closer look at a critical juncture, the transition separating the Neolithic from what used to be called the Bronze Age. What were the critical factors that propelled a sleepy village society from a stately *andante* to ever more rapid tempos?

War as an Agent of Cultural Change

From an evolutionary perspective war served as a kind of subset of the more general principle that one thing leads to another. Thus war leads to peace, and peace, once the dust settles, leads again to a new war.[1] Usually people learned something during each cycle, even if it was only how to make war more efficiently. When the cycle ended there had been progress, in the sense I defined it back in Variation I.[2]

War is such a recurrent theme in the vast tapestry of human history that we cannot dismiss it as just another flaw in human nature. Something else is at work here, some feature basic to the evolutionary track our species was following.[3]

Here are some of the more obvious reasons why war as-

---

1 Turchin, Peter. *War and Peace and War, The Rise and Fall of Empires.* (Penguin, 2007).

2 To save the reader trouble, my definition is restated here: "Progress is ... the direction followed by evolution. It relates to evolution the way decay and disintegration relate to entropy. It follows that progress happens when changes occur that move whatever is changing in the direction of concentration of power or greater complexity, or both."

3 Morris, Ian. "The Evolution of War." *Cliodynamics*, Vol. 3, Issue 1 (2012).

sumed the role it did in the early cultural evolution of our species:

- Whatever the proximate causes of the conflict, war usually resulted in the victors getting bigger, both in resources and in territory. As the groups grew in size, their capacity to provide for their citizens grew also. But problems of governance grew more complex as well. The emergence of new problems created new demands for new workarounds and new institutions.

- The victorious units were able to use slavery to add to their existing power base. Slaves could do jobs domesticated animals could not. They built roads and palaces, and almost everything else that required intensive labor. The more nubile females among the losers often ended up as concubines for the victors, providing a back door route to both genetic and cultural hybrid vigor.

- War provided an enormous stimulus for technological innovation. If your team lost a battle because the other team had bronze weapons and you didn't, you became powerfully motivated to acquire this new-fangled bronze technology yourself. This kind of arms race continues to this day.

- Wars were won or lost not only on weaponry but on organizational skills and discipline. Hierarchies and rules evolved rapidly and the lessons learned spilled over into civilian governance as well.

- War had a decisive effect on gender relationships. Patriarchal rule, and male dominance generally, can trace their origin not to innate human nature but to this period when wars were fought under circumstances where the male's body strength gave an advantage.

- Perhaps most important of all, wars provided powerful support for the development of group loyalty. Problems of governance multiply as the size and complexity of the unit increases. When concern for one's reputation isn't enough to keep the malingerers in line, the law steps in to fill the breach. When that is inadequate, and non-cooperation grows to the point that it threatens the integrity of the group, there's nothing like a good brisk war to get everyone marching in step. Fear of conquest by some alien force is a powerful motive for cooperation.

War didn't evolve by itself, any more than altruism did at an earlier stage in our evolution. One essential partner for war was religion, which co-evolved with it.

The Problem of Motivation

During the Neolithic era, increasingly complex hierarchies evolved separating commoners from the monarch and his community of aristocrats. When war came, the lords lorded over the whole procedure and reaped most of the benefits, while commoners did most of the fighting and suffered the heaviest casualties. What on earth motivated the common soldier to put his life on the line for someone else's cause?

The answer is that religion adapted to the new situation and became the prime motivator for soldiers when they faced the prospect of a sudden and bloody death. If you are about to die it helps if you believe in an afterworld. But there's more to it than that.

The great world religions of our modern era began during eras of social unrest and frequent war. They answered many other human needs, most of them with no connection to war at all, but they each embodied symbolism that could be invoked in time of war. "Onward, Christian Soldiers..." "...*Allahu akbar*...", and so forth. The effect was, and still can be, that of an aphrodisiac, particularly effective with younger males. It's still with us. It is the social equivalent of the adrenalin rush that an individual gets when facing sudden danger.

That adrenalin rush is part of our human nature. Its use as a facilitator for war is, however, a cultural adaptation. This means that war is *not* inevitable at all times and under all circumstances. It is a cultural artifact, not a genetically transmitted product of natural selection. We learned to live with war and we can learn to live without it.

The same goes for the gods of war we grew up with, Yahweh, Allah, and the more militant avatars of the Christian deity. We no longer need to see them up in the heavens someplace, "...trampling out the vintage where the grapes of wrath are stored." Indeed, we're a lot better off if we don't.

In Defense of Group Selection Theory

The rules of the selection game changed as war became endemic. Natural selection theory could cope with competition between species over a time span based on generations, but was unable to cope with this new phenomenon of intentional competition based on much shorter time spans that occurred within a single species.

Group selection theory is a better analytical tool than natural selection in coping with this new kind of competition because it works with shorter time spans and deals with intentional factors as well as natural ones.

Equally important, it can cope with adaptations that reverberate on more than one of the several hierarchical levels that evolved as human societies became more complicated. If a specific workaround or institutional wrinkle is introduced at a group level, and it also causes happiness at the personal level, it is more likely to evolve as a permanent feature of the institutional landscape.

This kind of serendipity can be relatively simple, as with the example of prayer, or it can be very complicated. The more complex the society, the more complicated the mechanisms and attitudes needed to reinforce the structural integrity required to keep the whole edifice from collapsing. We'll take a closer look at this presently.

SEVENTH VARIATION
THE DYNASTIC ERA

When a society grows it becomes more complicated; the attitudes and institutions that hold the group together become more complicated as well. It's a never-ending treadmill and just seems to keep on getting worse. Every time an old problem is solved, new ones pop up.

The problem is especially acute when the number of hierarchies increases. To wage war effectively, you need a line of command, from generals to privates. Managing an empire also can require many levels, each with its own problems and challenges.

Hierarchies, like buildings, have to be built from the ground up. If you try to construct higher levels before the intermediate ones are firmly in place you are in trouble, as some of America's recent efforts at nation-building abroad have discovered.

It's like adding new floors to a multistory building that needs structural integrity and other new arrangements between floors, as well as within each floor. Add one floor and you need stairs, plus you may need to look to your foundations. Add a couple more floors and you'll have to look to your foundations and materials, plus you may want an intercom, fire escapes and an elevator. And so on.

We work at resolving all these problems as they arise because of the considerable benefits that accrue to the individuals who succeed in coming together in larger groups. We look ahead and wonder why we bother. We look back and we know.

Who's in Charge?

Every organized group of people needs someone with the authority to make decisions for the group as a whole. Leadership at the village level evolved during the Neolithic and beyond into ever more elaborate systems of top-down governance. The task of holding increasingly large and complex societies together while they digested the spoils of war and the fruits of technological advance was not an easy one, and got more demanding with the passage of time.

Rulers formed their own "in-group" (the aristocracy) and commoners had separate ones for their own communities. Mediating between the two levels were separate groups of specialists like priests and tax collectors. My, how these in-between groups have proliferated since those early beginnings!

The problem of what to do when a leader dies became acute when Neolithic villages began to coalesce into larger units. It's not so difficult when everyone knows everyone else, but when a society grows and develops segments some institutional mechanism that is generally recognized has to be in place. Otherwise, when a leader dies, nobody knows which of the leaders of the various smaller units should take over.

It was only natural (kin-based altruism) that at first the need for such a mechanism was met by the principle of primogeniture. The eldest child of the ruler took over when the old leader died. This was sufficiently clear-cut so that all who accepted the principle knew who the next ruler would be. Almost everyone did accept it, for a long time, ushering in what you could call the dynastic era in human history.

Since this was also the era when war played a central role in the evolutionary process, kings and emperors were almost always male, with the line of succession through male offspring only.

Kingdoms and Empires

The dynastic system's greatest success was in making possible a new kind of society within which several subtribes could cooperate symbiotically. This required an enlarged sense of group identity, something that would provide for altruistic behavior toward other members of the group, even if they belonged to other subtribes within the system and even if they were relative strangers. If it was to hang together, the larger group needed ways to enlarge and fortify the ancient "us versus them" sense that people had inherited from a simpler, more natural era. Dialects, appearance, ethical codes, and religion all blossomed in response to this need.[1]

Early on, dynastic societies began the march toward what we call civilization. They managed a pretty long run, but within the last several centuries they have bumped against a ceiling beyond which they could achieve little further progress. Part of it was energy—you can only do so much if you depend on animals and slaves and wind to power your lives. But most of that ceiling was the enormously complicated nature of the societies that had evolved under hereditary leaders. Further evolution required basic political as well as technical changes.

Dynastic systems have always suffered from flaws associated with primogeniture. Systems based on a hereditary monarch who actually exercises power himself usually run out of steam after a couple of generations. The crown prince ascends

---

1 Coon, Carl. *One Planet, One People*. (Prometheus, 2004). See especially Chapter 10 for a more detailed analysis.

to the throne and it soon turns out that he lacks the unusual qualities of character and intellect that enabled his father to get the enterprise going in the first place. If the system the father established is a solid one, and the son is reasonably competent, the dynasty may rock along for a while and even flourish. But when the grandson, who was pampered from childhood, takes over, he may well botch things up, perhaps irretrievably. Or he may become a figurehead with other, tougher individuals managing the show behind the scenes.

One example of how the dynastic way of life runs its course is the way the princely states fared in the Indian subcontinent. A scant three hundred years ago most of the maharajas and rajas in the princely states had absolute authority within their own territories.[2] The British Raj infiltrated the system through a combination of bribery, trickery, and military force, and pretty soon almost all of these titular monarchs were figureheads, with a British agent at their elbows calling the shots. What little authority they retained was lost when an independent Indian government bought them out, putting them on a stipend.

The French and Russian revolutions are examples of a more abrupt kind of transition. But whether the old regimes went out with a bang or a whimper, they were doomed not so much by the incompetence of their royal leaders (though that helped) as by the evolution of the societies they managed to levels of complexity that required a new form of governance.

One trigger that paved the way for a new political environment was the development of fossil fuels as a major source of power.

---

2 It's an interesting commentary on the inherent strength of the village-sized community that during the heyday of the dynastic era on the Indian subcontinent, the crowned heads and their families bonded with each other, across the bounds of the various political units, both socially and for marriages. This was also true to some extent in Europe.

EIGHTH VARIATION
THE MODERN ERA

When you graph global population increase over the last 10,000 years you see what some economists call a hockey stick. That is, the line stays fairly flat, close to zero, for the first three quarters of your graph, and then begins to rise. Starting during the eighteenth century the angle increases, and the line continues to rise ever more steeply as we approach the present.

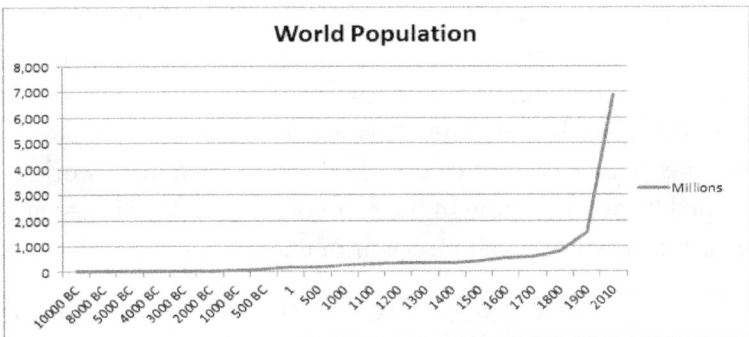

The accelerating upward trend this graph[1] shows is partly due to increasingly sophisticated governance, that evolved as people graduated from the old-fashioned dynastic systems, but it also reflects a huge augmentation in the ability to concentrate energy.

The breakthrough on the energy front came with the switch to fossil fuels as the main source of energy. A good place to start is with the invention of the steam engine in England. It was initially powered by wood but when the forests in England became depleted people turned to coal, which opened up a host of further possibilities. Pretty soon there were coal-powered steam locomotives and steamships.[2] Then petroleum entered the picture, and finally, nuclear power. In our time, many people are learning to harness solar energy more efficiently, and some are tinkering with fusion power. If they ever learn to harness that we'll achieve a state where power is as plentiful as air and water.

If evolution is defined by concentration of energy as well as by the emergence of complexity, this exponential increase in our access to power over the relatively short period of a quarter of a millennium has to be ranked among the major events in the evolution of evolution itself.

The hockey stick metaphor can also be used for looking at the rise in the complexity of our social institutions. When the American colonies said goodbye to the British, they resolved to establish a new system of governance that rejected the whole idea of monarchy and renounced the old buddy system between church and state. This was the first major instance of an explicit rejection of the old dynastic system. It worked, helped by good fortune in the form of an equable climate and a vast, nearly untapped reservoir of natural resources. Its suc-

---

1 Data sources: *United States Census Bureau - Historical Estimates of World Population; United States Census Bureau -Total Midyear Population for the World, 1950-2050*

2 Morris, Ian. *Why the West Rules—For Now.* (New York: Random House, 2011), pp. 498 ff.

cess fired the imaginations of political activists over much of the rest of the world, and very soon (in terms of earlier evolutionary tempos) old fashioned governments were being replaced by governments that at least said they were constitutional democracies or monarchies.

On the plus side, slavery was abolished, technology helped bring about an increase in food supplies, and population shot up. The old patriarchal model was shredded. Women demanded equal rights and now, with warfare increasingly a technical matter rather than one depending on brute strength, they are beginning to get them. Modern transportation is destroying spatial barriers and people are mixing as never before, eroding old regional and racial differences. The information revolution that just started is transforming not only the way we communicate, but the way we think.

On the down side, we've incurred a whole cascade of new problems. Abandoning the old dynasties for new and relatively untested models brought vast new problems of governance to the fore. By now the older democracies, with generations of experience in handling these problems, have varying degrees of success, but much of the world is still having acute growing pains. We now talk about failed states, as well as developing ones. Regional wars are still endemic, and the threat of a big global war, risking nuclear Armageddon, still hangs over us.

The problems we still face usually involve how to ensure cooperation between groups that have not yet learned to work together. The new structures, to fall back on our architectural analogy, require new systems of interoffice communication, new structural materials to reduce costs and improve performance, and many other innovations. Some of these requirements have been solved and others have temporary fixes, but many still remain testimony to the unfinished business of the new environment humanity created with the industrial revolution.

There is no point in trying to itemize the problems we face and describe our efforts to manage them. Any university course catalogue will do better. Over half of all the courses offered will be responses to meet these new challenges, whether it is, for example, political science or economics or law or business administration. We are at best only halfway toward success in assimilating the fruits of our newly gained access to power.

To cap all this, climate change is looming. Maybe that's what our perplexed and bothered species needs at this point to persuade us to lift our sights and cooperate for the common good as a single species. We have become too efficient at waging war to have a good brisk one that is serious enough to scare the countries on the cutting edge enough to make them start pulling together. Climate change may be coming back to center stage to replace war, as war replaced it ten thousand years ago. There is a certain Shakespearean quality to this sequence.

The central feature of our times is that in everything that counts, we have almost won the race to a very important ceiling. Although governance problems remain acute, we now have workarounds to the point that the infrastructure of global governance is in place. We've pushed the frontiers of our knowledge out into the cosmos and down to the depths of subatomic particles. Meanwhile, we've come a long way towards gender equality and are busily engaged in reestablishing the essential unity of our species, culturally and even to some extent genetically. We are beginning to learn a bit more about ourselves, and how we relate to each other.

We still have problems and we won't solve all of them right away, but right now we are not only pushing on a brand new ceiling, we are knocking a few cracks in it, enough to see the light on the other side, if we know where to look.

NINTH VARIATION
MORALITY

Our "short history" outlines the case for the scientific view that life and human civilization evolved without divine intervention. With evolution as our central theme, we identified a chain of events that leads from the primordial soup to my sitting here writing this explanation. The evidence for most of these events comes from such subjects as astronomy, cell research, climatology, paleoarcheology, and of course the historical record.

In this chapter we shall discuss certain moral and philosophical issues that have been bothering thoughtful people for thousands of years. We cannot possibly try to summarize the substance of their arguments and conclusions in this short narrative, but I hope to demonstrate that our understanding of evolution can provide fresh perspectives on some of them.

The Origin and Evolution of Morality

We can infer that our remote ancestors began to develop a human sense of right and wrong behavior after they left the jungles of Southern Africa and learned to survive as bipedal hunters. The change to monogamy goes back to this period as far as we know, and so does the practice of sharing kills of the larger mammals with the group as a whole, rather than just feeding the immediate family. These behavioral changes must have evolved symbiotically with the physical changes the archeological record shows for that ancient era when the switch to savannah life occurred.

Much later, *erectus* and its cousins developed fire and, here and there, the first signs of ritual burial. However, morality as we know it probably remained only latent until *sapiens* arrived on the scene, and language evolved along with altruism, making possible the evolution of larger societies. When reciprocity and reputation entered the equation concepts were needed to describe whether individuals could be trusted. Morality provided those concepts, with the concepts of good and evil as standards for judging behavior.

The concepts of good and evil proved useful for a lot more than gossip within the group, when groups became larger and more complex. Simple village level ways of judging others didn't work too well at the tribal level and hardly worked at all when wars produced empires and people dealt regularly with strangers. Concepts of right and wrong grew more complex and were supplemented by legal systems and other means of coercion. Religion emerged, and then nationalism, to support the new multi-tiered societal structures.

Morality, in short, evolved along with other features of cultural evolution, not as an isolated phenomenon. It was both an enabler of other kinds of evolution and was enabled by them.

Human Nature

Human nature as generally understood covers behavior that is innate, not learned. You cannot change how your genes are encoded, though you can usually override their instructions as necessary—this is a large part of what civilization is all about. But when learning something "comes natural" the whole learning process becomes quicker, easier, and more durable.

The accuracy with which genes transmit information is now well established and is a foundational concept for our understanding of evolution. We define life itself in terms of genetic descent. Nevertheless, where we used to see a sharp line between behavior that is transmitted genetically and that which is learned, or between nature and nurture in common parlance, we now see a zone, an area that exists in between like the intertidal area between land and sea. There is a good bit of research going on at present into how this zone between nature and nurture works.1

Biological evolution proceeds at a much slower tempo than cultural evolution, and efforts to change human behavior take a lot longer when that behavior is part of human nature. Most of us are fully aware of this distinction when applying it to individuals, but less so when dealing with whole societies. Mao Tse-tung tried to alter the human nature of a whole nation in a generation, with catastrophic results.

Salvation and Original Sin

Any good stew starts with solid ingredients. When paternalistic, belligerent, war-prone religions evolved out of the Neolithic era, the old recipes weren't enough. One of the new ingredients was the idea of salvation, which worked by embellishing the ancient belief in life after death with the proviso that you have to earn your ticket by following the rules before you can go to heaven.

One way to spice up the idea of salvation was to add the idea of redemption. But who needs redemption and why? What is there to redeem? One answer is the idea of original sin, which tells us that we have to work at redemption, for

1 Jablonka, Eva and Lamb, Marion. *Evolution in Four Dimensions.* (MIT Press, 2005); Carey, Nessa. *The Epigenetics Revolution.* (Icon Books, 2013).

if you're born a sinner you can't go to heaven automatically, you have to earn it. This rather harsh view of human nature was seized on by Christians, who made it a central pillar of their faith. This idea persists today, largely within Christian denominations.

The concept of original sin has no place in humanist thought. We do not believe a child is born guilty of anything. There are more humane ways of encouraging responsible social behavior.

Free Will, Infinity, and Divinity

All through this narrative we've been dealing with issues of scale. To get even an inkling of what was going on when the first life on earth began, we had to think in terms of hundreds of millions of years. But that scale was useless when we looked for clues to explain the origin of our species. There we had to pull back our zoom lens and survey the scene in terms of hundreds of thousands of years. A scale of thousands of years served for the Paleolithic and perhaps the early Neolithic eras, and after that we started measuring history in centuries. Now we have the daily news.

The same principle applies to space. We need a very different focal lens to view atomic particles from the ones we use to peer outward at galaxies, and in between there is a small slice of the space continuum that constitutes the world we grew up in.

Scientists help us extend our reach from atoms to galaxies and from micro-milliseconds to eons, but for most of us, the small slice of time and space we inhabit is enough. We are preoccupied with our own time/space bubble, and while explorations outside it can be interesting and have sometimes proved useful, we usually leave them to the specialists.

Nevertheless, the concept of infinity has always troubled us. If the world we live in is composed of things that result from prior causes, which in turn were caused by something else and so on, can you go back in time and space to a first cause or do you just keep on going on, and where does it all end? This is simply one way to express the basic paradox of infinity, which forms the basis for the issue of free will versus determinism. Am I responsible for actions which were caused by other events in a chain of causation that goes back indefinitely?

One conclusion that emerges from our study of evolution is that humans are most likely the first and only species that worries about this issue. The paradox that lies at its core is a product of the human capacity for abstract thought, not something that actually exists in the space-time continuum we inhabit.

We might conclude, on this basis, that it isn't important. But it is, because perplexity over infinity leads us all too often to belief in divinity. We see things happening that we now know can be explained but only with different lenses on our camera, and those lacking those lenses usually invoke supernatural causes. Galileo got in trouble with the Jesuits partly because he questioned dogma about the nature of infinitely small entities.[2] Most people still believe that if you keep pushing long enough, you'll eventually get to some root cause. Since they don't know what else to call it, they call it divine.

Nothing we have learned equips us to answer the paradox embodied in the idea of infinity, but we now can put it in better perspective. If we can consider free will and divinity as issues arising from the paradox, and see them in an evolutionary context, we can conclude that the paradox itself is insoluble and can stop worrying about it. We have plenty of other issues that need our attention more urgently.

---

[2] Alexander, Amir. *Infinitesimal: How a Dangerous Mathematical Theory Shaped the Modern World.* (New York: Farrar, Straus and Giroux, 2014).

CODA

Recapitulation

The last several chapters brought our narrative up to the present. That would probably be a good place to quit, since predicting the future is a very chancy business. When the subject is life itself, and our species in particular, there are just too many variables.

On the other hand, we've made a couple of assumptions about the essential nature of the evolutionary process. If they have helped us to understand what we know about the past, maybe they also can give us clues about what might happen in the future. If the shoe fits, wear it.

Our first assumption was that evolution is a universal principle, like gravity or entropy. The second was that evolution operates to produce complexity out of relative simplicity.

The third is that in this process energy becomes concentrated. That is, the power available to the entity that is evolving tends to increase.

The trends toward complexity and concentration of power tend to fade out of sight when we look at what is evolving with the wrong time scale. If we look at the collapse of the Roman Empire in terms of centuries the trend is toward less complexity and less concentration of power, not more. But when you look at the evolution of Western civilization from the Neolithic era up to the present, the trend is easier to see.

The evolutionary process works because things that work better tend to last longer. But the process of winnowing out what works best and discarding the dross can be chancy and time-consuming. Progress comes in fits and starts, and often appears to involve several steps back as well as some steps forward. But there does seem to be something like a ratchet principle that operates over the long run to ensure that if you wait long enough, there will be progress. That's really what evolution is all about.

Over longer time spans, progress is punctuated by what I call ceilings. Call them parameters, or algorithms, or whatever fancy names you will, they define the rules of the game within which an evolutionary process plays itself out during a finite slice of time and space. As that game continues and perfection within those rules approaches, pressures mount for some game-changing breakthrough that will open up possibilities for major new advances. Seen close up, such developments can seem enormously important, but viewed from afar they are just blips on a screen. Think of them as ratchets writ large.

The Perfect Storm

By many standards, we live in interesting times. Several major evolutionary forces are converging, all bumping against ceilings, all demanding some kind of breakthrough. It is like a perfect storm.

What are these "rare combinations of circumstances"? Well, on the energy front, it's becoming clear we need to change our ways. We still have substantial reserves of fossil fuels, but our needs for power to fuel a rapidly growing world economy are growing. The energy breakthrough has to come in the form of renewable energy sources, and the need may become more urgent sooner than most of us expect. The climate is beginning to change and no matter who is to blame our needs for power will drastically increase. Meeting those needs with clean power is one of the principal challenges we face.

Meanwhile the population bomb has already exploded. We cannot just keep on expanding out into new turf as in the past because the earth's surface is finite and we are coming close to filling it.

On the complexity front, the central issue for our species is governance, how to build systems that bring ever larger groups of individuals into cooperative relationships. There are really two issues here, how to keep the peace in the more quarrelsome neighborhoods, where people still believe war is the answer, and how to create some central authority that can govern on issues of global concern. It seems to me that on both counts we need a stronger United Nations, one that has both the will and the capacity to intervene decisively in the world's worst conflicts, while providing the platform on which effective global governance for all of us can evolve.

A key sticking point is that cooperation between the major powers can no longer be secured by the classic expedient of fighting it out, because with the nukes, war has become just too expensive. Somehow, we have to find another way to global governance. This is new ground we're exploring, and we aren't there yet by a long shot.

Where Do We Go From Here?

But how do we get from here to there? The United States is a natural fit for the type of leadership the rest of the world needs, but hubris has led us down the path of believing that the UN be damned, we shall fulfill the role of global sheriff ourselves. Meanwhile, we haven't even sorted out our own problems of internal governance, so we fiddle while an increasingly impatient world watches and considers alternatives.

The patterns of human societal evolution we have observed since the Neolithic era suggest two possibilities. The first is a reversion to dynastic principles for the time being, with global rule by oligarchy rather than a single titular monarch. This might provide the central authority humanity needs to get people cooperating enough to resolve some of our more acute global problems. I suspect some of the billionaires that hobnob in Davos, and some others who are busy buying elections in our country, would be willing to shoulder the responsibilities of being the world's policeman as and when they are able to grab enough power.

This might prove an expedient way of getting our quarrelsome species past some of our more acute present migraines, but pretty soon the old problem of how authority gets passed on between generations would lead to a new era of disorder. At least we'd have learned something (the ratchet principle) that would pave the way for the eventual evolution of a more durable form of global government.

The other path forward would lead directly to a form of global governance grounded on the world's recent experiences with representative government and inspired by a deep respect for humanity as a whole. That, of course, is eminently the more desirable alternative.

Something has to give. We cannot just stay put where we find ourselves now, in a halfway house between a world dominated by competing nations and a global authority. Either we go forward or, like Rome, we descend into a long period of relative darkness. Perhaps that too would constitute just another blip on the history of evolution, but even if seen from a long way off, it would be an unusually big one, and very painful for the people living at the time.

Frankly, I find it's hard to be optimistic when I read the daily press about the latest happenings in Washington or at UN headquarters. But who knows, evolution has a way of asserting itself, even when the short run obstacles look overwhelming.

Let's all keep looking for cracks in that ceiling we're bumping against and catch some glimpses of the bright future that awaits us if we can just get through to it. If enough of us can sense that bright future we may achieve it sooner rather than later.

Carl Coon
April, 2014

Picture Credits

Page 7 (Darwin sitting): [public domain]

Page 10 (Darwin's finchews): [public domain]

Page 19 (homeless man): Image credit: halfpoint / 123RF Stock Photo

Page 31 (cave painting): Ancient cave paintings in Patagonia. Image credit: Adurivero / 123RF Stock Photo

Page 37 (wheat): Image credit: kobyakov / 123RF Stock Photo

Page 42 (war painting): Image credit: lolloj / 123RF Stock Photo

Page 47 (crown): Image credit: rorem / 123RF Stock Photo

Page 51 (geothermal power plant): [public domain]

Page 55 (compass): Image credit: paulfleet / 123RF Stock Photo

Page 60 (brain gears): Image credit: lightwise / 123RF Stock Photo